Types of Rocket Propulsion and Potential Space Drives

There is lots of interest these days in space exploration. After fifty years we are looking at going back to the Moon to build a base and early planning for a manned trip to Mars.

After that, further exploration and settlement of the Solar System will be on the table.

One of the main limiting factors for all of this space exploration are the types of rockets and rocket engines available to get there.

Current rockets in usage are mainly chemical rockets with fairly low specific impulses. We will need a lot more powerful options to reach all of these destinations.

This book is about the history of rocket engine development, what we are using today, and some options for the future.

We also discuss some exciting thinking about how we could build interstellar and faster than light space propulsion for exploring the nearest stars too.

Types of Rocket Propulsion and Potential Space Drives

Types of Rocket Propulsion and Potential Space Drives

Types of Rocket Propulsion and Potential Space Drives

Types of Rocket Propulsion and Potential Space Drives

Other books by Martin K. Ettington

Spiritual and Metaphysics Books:
Prophecy: A History and How to
Guide
God Like Powers and Abilities
Enlightenment for Newbies
Removing Illusions to Find True
Happiness
Using the Scientific Method to
Study the Paranormal
A Compendium of Metaphysics
and How to Guides (Six
books together in one
volume)
Love from the Heart
The Enlightenment Experience
Learn Your Soul's Purpose
Pursuing Enlightenment
A Modern Man's Search for Truth
Use Intuition and Prophecy to
Improve Your Life
The Handbook of Spiritual and
Energy Healing

Longevity & Immortality:
Physical Immortality: A History
and How to Guide
The Commentaries of Living
Immortals
Records of Extremely Long Lived
Persons
Enlightenment and Immortality
Longevity Improvements from
Science
The 10 Principles of Personal
Longevity
Telomeres & Longevity
The Diets and Lifestyles of the
Worlds Oldest Peoples
The Longevity Six Books Bundle

Science Fiction:
Out of This Universe
Personal Freedom-Parts 1 & 2
The Psychic Soldier Series:
    Book 1-Himalayan Journey
    Book 2-A Soldier is Born
    Book 3-Fighting For Right
    Book 4-Earth Protector
The Immortality Sci Fi Bundle

The God Like Powers Series:
Human Invisibility
Invulnerability and Shielding
Teleportation
Psychokinesis
Our Energy Body, Auras, and
Thoughtforms

The God Like Powers Series—
Volume 1 Compilation
The Yoga Discovery Series:
Yoga-An Ancient Art Form
Hatha Yoga-Helping you Live
Better
Raja Yoga-Through the Ages
The Yoga Discovery Package

Business & Coaching Books:
Creating, Paublishing, &
Marketing Practitioner
Ebooks
Building a Successful Longevity
Coaching Business
Why Become a Coach?
The Professional Coaching
Success Trilogy
2020-Make Money Writing and
Selling Books
The 2020 Handbook of High
Paying Work Without a
College Degree

Science, Technology, and Misc.
Future Predictions By and
Engineer & Seer
The Unusual Science &
Technology Bundle
The Real Atlantis-In the Eye of
the Sahara
Are Cryptozoological Animals
Real or Imaginary?
Real Time Travel Stories From a
Psychic Engineer
Removing Limits On Our
Consciousness-And
Thinking Outside the
Box

Types of Rocket Propulsion and Potential Space Drives

These books are all available in digital and printed formats from my website and on Amazon, Barnes & Noble, Apple ITunes, and many other sites

My Books Website is: http://mkettingtonbooks.com

Types of Rocket Propulsion and Potential Space Drives

Signup for our Mailing List to get the following:

1) A discount coupon for 25% discount on all books on our site

2) Occasional Notices of new books available

3) Occasional Email on other offerings of ours (Monthly)

Go to this link to sign-up:

http://personal-longevity.com/mkebooks/emailsignup/

And click this link to get the FREE 102 page Ebook titled "Secrets of Many Things"

If you have any questions about this book or other subjects please contact the Author at:

mke@mkettingtonbooks.com

Types of Rocket Propulsion and Potential Space Drives

Types of Rocket Propulsion and Potential Space Drives

Table of Contents

Types of Rocket Propulsion and Potential Space Drives

Types of Rocket Propulsion and Potential Space Drives

1.0 Introduction

There is lots of interest these days in space exploration. After fifty years we are looking at going back to the Moon to build a base and early planning for a manned trip to Mars.

After that, further exploration and settlement of the Solar System will be on the table.

One of the main limiting factors for all of this space exploration are the types of rockets and rocket engines available to get there.

Current rockets in usage are mainly chemical rockets with fairly low specific impulses. We will need a lot more powerful options to reach all of these destinations.

This book is about the history of rocket engine development, what we are using today, and some options for the future.

We also discuss some exciting thinking about how we could build interstellar and faster than light space propulsion for exploring the nearest stars too.

Types of Rocket Propulsion and Potential Space Drives

2.0 Definition of Terms

When talking about rocket engines and launch vehicles there are several important measurements to keep in mind to understand the differences between the designs. Here are some of those terms:

Specific Impulse

Specific impulse (usually abbreviated Isp) is a measure of how effectively a rocket uses propellant or a jet engine uses fuel. Specific impulse can be calculated in a variety of different ways with different units. By definition, it is the total impulse (or change in momentum) delivered per unit of propellant consumed and is dimensionally equivalent to the generated thrust divided by the propellant mass flow rate or weight flow rate. If mass (kilogram, pound-mass, or slug) is used as the unit of propellant, then specific impulse has units of velocity. If weight (newton or pound-force) is used instead, then specific impulse has units of time (seconds). Multiplying flow rate by the standard gravity ($g0$) converts specific impulse from the weight basis to the mass basis.

A propulsion system with a higher specific impulse uses the mass of the propellant more efficiently. In the case of a rocket or other vehicle governed by the Tsiolkovsky rocket equation, this means less propellant needed for a given delta-v. In rockets, this means that the vehicle the engine is attached to can more efficiently gain altitude and velocity. This effectiveness is less important in jet aircraft that use ambient air for combustion, and carry payloads that are much heavier than the propellant.

Specific impulse can include the contribution to impulse provided by external air that has been used for combustion and is exhausted with the spent propellant. Jet engines use outside air, and therefore have a much higher specific impulse than rocket engines. The specific impulse in terms of propellant mass spent has units of distance per time, which is a notional velocity called the effective exhaust velocity. This is higher than the actual exhaust velocity because the mass of the combustion air is not being accounted for. Actual and effective exhaust velocity are the same in rocket engines operating in a vacuum.

Specific impulse is inversely proportional to specific fuel consumption (SFC) by the relationship Isp = 1/(go·SFC) for SFC in kg/(N·s) and Isp = 3600/SFC for SFC in lb/(lbf·hr).

Payload to LEO

This terms applies to how much weight the rocket can launch into Low Earth Orbit or LEO. The largest Payload rocket to date was the Saturn Five used in the Apollo program to land man on the Moon.

The Saturn five payload for LEO was 310,000 pounds or 140,000 kilograms.

3.0 The History of Rockets

According to the writings of the Roman Aulus Gellius, the earliest known example of jet propulsion was in c. 400 BC, when a Greek Pythagorean named Archytas, propelled a wooden bird along wires using steam. However, it would not appear to have been powerful enough to take off under its own thrust.

The aeolipile described in the first century BC (often known as Hero's engine) essentially consists of a steam rocket on a bearing. It was created almost two millennia before the Industrial Revolution but the principles behind it were not well understood, and its full potential was not realized for a millennium.

The availability of black powder to propel projectiles was a precursor to the development of the first solid rocket. Ninth Century Chinese Taoist alchemists discovered black powder in a search for the elixir of life; this accidental discovery led to fire arrows which were the first rocket engines to leave the ground.

It is stated that "the reactive forces of incendiaries were probably not applied to the propulsion of projectiles prior to the 13th century". A turning point in rocket technology emerged with a short manuscript entitled Liber Ignium ad Comburendos Hostes (abbreviated as The Book of Fires). The manuscript is composed of recipes for creating incendiary weapons from the mid-eighth to the end of the thirteenth centuries—two of which are rockets. The first recipe calls for one part of colophonium and sulfur added to six parts of saltpeter (potassium nitrate) dissolved in laurel oil, then inserted into hollow wood and lit to "fly away suddenly to whatever place you wish and burn up everything". The second recipe combines one pound of sulfur, two pounds of charcoal, and six pounds of saltpeter—all finely powdered on a marble slab. This powder mixture is packed firmly into a long and narrow case. The introduction of saltpeter into pyrotechnic mixtures connected the shift from hurled Greek fire into self-propelled rocketry. .

Articles and books on the subject of rocketry appeared increasingly from the fifteenth through seventeenth centuries. In the sixteenth century, German military engineer Conrad Haas (1509–1576) wrote a manuscript which introduced the construction to multi-staged rockets.

Rocket engines were also brought in use by Tippu Sultan, the king of Mysore. These rockets could be of various sizes, but usually consisted of a tube of soft hammered iron about 8 in (20 cm) long and 1 1/2–3 in (3.8–7.6 cm) diameter, closed at one end and strapped to a shaft of bamboo about 4 ft (120 cm) long. The iron tube acted as a combustion chamber and contained well packed black powder propellant. A rocket carrying about one pound of powder could travel almost 1,000 yards (910 m). These 'rockets', fitted with swords, would travel long distances, several meters in the air, before coming down with swords edges facing the enemy. These rockets were used very effectively against the British empire.

Modern rocketry

Slow development of this technology continued up to the later 19th century, when Russian Konstantin Tsiolkovsky first wrote about liquid-fueled rocket engines. He was the first to develop the Tsiolkovsky rocket equation, though it was not published widely for some years.

The modern solid- and liquid-fueled engines became realities early in the 20th century, thanks to the American physicist Robert Goddard. Goddard was the first to use a De Laval nozzle on a solid-propellant (gunpowder) rocket engine, doubling the thrust and increasing the efficiency by a factor of about twenty-five. This was the birth of the modern rocket engine. He calculated from his independently-derived rocket equation that a reasonably sized rocket, using solid fuel, could place a one-pound payload on the Moon.

The era of liquid fuel rocket engines

Goddard began to use liquid propellants in 1921, and in 1926 became the first to launch a liquid-propellant rocket. Goddard pioneered the use of the De Laval nozzle, lightweight propellant tanks, small light turbopumps, thrust vectoring, the smoothly-throttled liquid fuel engine, regenerative cooling, and curtain cooling.

During the late 1930s, German scientists, such as Wernher von Braun and Hellmuth Walter, investigated installing liquid-fueled rockets in military aircraft (Heinkel He 112, He 111, He 176 and Messerschmitt Me 163).

The turbopump was employed by German scientists in World War II. Until then cooling the nozzle had been problematic, and the A4 ballistic missile used diluted alcohol for the fuel, which reduced the combustion temperature sufficiently.

Staged combustion was first proposed by Alexey Isaev in 1949. The first staged combustion engine was the S1.5400 used in the Soviet planetary rocket, designed by Melnikov, a former assistant to Isaev. About the same time (1959), Nikolai Kuznetsov began work on the closed cycle engine NK-9 for Korolev's orbital ICBM, GR-1. Kuznetsov later evolved that design into the NK-15 and NK-33 engines for the unsuccessful Lunar N1 rocket.

In the West, the first laboratory staged-combustion test engine was built in Germany in 1963, by Ludwig Boelkow.

Hydrogen peroxide / kerosene fueled engines such as the British Gamma of the 1950s used a closed-cycle process (arguably not staged combustion, but that's mostly a question of semantics) by catalytically decomposing the peroxide to drive turbines before combustion with the kerosene in the combustion chamber proper. This gave the efficiency advantages of staged combustion, whilst avoiding the major engineering problems.

Liquid hydrogen engines were first successfully developed in America, the RL-10 engine first flew in 1962. Hydrogen engines were used as part of the Apollo program; the liquid hydrogen fuel giving a rather lower stage mass and thus reducing the overall size and cost of the vehicle.

Types of Rocket Propulsion and Potential Space Drives

Most engines on one rocket flight was 44 set by NASA in 2016 on a Black Brant.

Types of Rocket Propulsion and Potential Space Drives

Types of Rocket Propulsion and Potential Space Drives

4.0 Major Historical Rockets & Engines

Rocket develop took off in the twentieth century. Robert Goddard pioneered liquid fueled rockets in the 1920s-1930s. And the Nazis developed an amazing leap in technology in the V-2 rocket even though it was designed to kill a larger number of civilians; it was still a major advance for rocketry at that time.

4.1 The German V2

The V-2 (German: Vergeltungswaffe 2, "Retribution Weapon 2"), with the technical name Aggregat 4 (A4), was the world's first long-range guided ballistic missile. The missile, powered by a liquid-propellant rocket engine, was developed during the Second World War in Germany as a "vengeance weapon", assigned to attack Allied cities as retaliation for the Allied bombings against German cities. The V-2 rocket also became the first artificial object to travel into space by crossing the Kármán line with the vertical launch of MW 18014 on 20 June 1944.

Research into military use of long-range rockets began when the studies of graduate student Wernher von Braun attracted the attention of the German Army. A series of prototypes culminated in the A-4, which went to war as the V-2. Beginning in September 1944, over 3,000 V-2s were launched by the German Wehrmacht against Allied targets, first London and later Antwerp and Liège. According to a 2011 BBC documentary, the attacks from V-2s resulted in the deaths of an estimated 9,000 civilians and military personnel, and a further 12,000 forced laborers and concentration camp prisoners died as a result of their forced participation in the production of the weapons.

As Germany collapsed, teams from the Allied forces—the United States, the United Kingdom, and the Soviet Union—raced to capture key German manufacturing sites and technology. Von Braun and over 100 key V-2 personnel surrendered to the Americans and many of the original V-2 team ended up working at the Redstone Arsenal. The US also captured enough V-2 hardware to build approximately 80 of the missiles. The Soviets gained possession of the V-2 manufacturing facilities after the war, re-established V-2 production, and moved it to the Soviet Union.

4.2 The Saturn V Moon Rocket

Saturn V was an American super heavy-lift launch vehicle certified for human-rating used by NASA between 1967 and 1973. It consisted of three stages, each fueled by liquid propellants. It was developed to support the Apollo program for human exploration of the Moon and was later used to launch Skylab, the first American space station.

The Saturn V was launched 13 times from Kennedy Space Center with no loss of crew or payload. As of 2020, the Saturn V remains the tallest, heaviest, and most powerful (highest total impulse) rocket ever brought to operational status, and holds records for the heaviest payload launched and largest payload capacity to low Earth orbit (LEO) of 310,000 lb (140,000 kg), which included the third stage and unburned propellant needed to send the Apollo command and service module and Lunar Module to the Moon.

Types of Rocket Propulsion and Potential Space Drives

As the largest production model of the Saturn family of rockets, the Saturn V was designed under the direction of Wernher von Braun at the Marshall Space Flight Center in Huntsville, Alabama, with Boeing, North American Aviation, Douglas Aircraft Company, and IBM as the lead contractors.

To date, the Saturn V remains the only launch vehicle to carry humans beyond low Earth orbit. A total of 15 flight-capable vehicles were built, but only 13 were flown. An additional three vehicles were built for ground testing purposes. A total of 24 astronauts were launched to the Moon in the four years spanning December 1968 through December 1972.

Saturn V Stages and Engines

The Saturn V consisted of three stages—the S-IC first stage, S-II second stage and the S-IVB third stage—and the instrument unit. All three stages used liquid oxygen (LOX) as the oxidizer. The first stage used RP-1 for fuel, while the second and third stages used liquid hydrogen (LH2). Whereas LH2 has a much higher energy density to be lifted into orbit by mass, RP-1 has a much higher energy density by volume. Consequently, RP-1 was chosen for the first stage propellant because the volume of LH2 required would have been more than three times greater and would have created much higher aerodynamic drag during the boost phase through the atmosphere. The upper stages also used small solid-propellant ullage motors that helped to separate the stages during the launch, and to ensure that the liquid propellants were in a proper position to be drawn into the pumps.

Types of Rocket Propulsion and Potential Space Drives

The S-IC was built by the Boeing Company at the Micloud Assembly Facility, New Orleans, where the Space Shuttle external tanks would later be built by Lockheed Martin. Most of its mass at launch was propellant: RP-1 fuel with liquid oxygen as the oxidizer. It was 138 feet (42 m) tall and 33 feet (10 m) in diameter, and provided over 7,600,000 pounds-force (34,000 kN) of thrust. The S-IC stage had a dry weight of about 289,000 pounds (131 metric tons); when fully fueled at launch, it had a total weight of 5,100,000 pounds (2,300 metric tons). It was powered by five Rocketdyne F-1 engines arrayed in a quincunx. The center engine was held in a fixed position, while the four outer engines could be hydraulically turned with gimbals to steer the rocket. In flight, the center engine was turned off about 26 seconds earlier than the outboard engines to limit acceleration. During launch, the S-IC fired its engines for 168 seconds (ignition occurred about 8.9 seconds before liftoff) and at engine cutoff, the vehicle was at an altitude of about 36 nautical miles (67 km), was downrange about 50 nautical miles (93 km), and was moving about 7,500 feet per second (2,300 m/s).

S-II second stage

The S-II was built by North American Aviation at Seal Beach, California. Using liquid hydrogen and liquid oxygen, it had five Rocketdyne J-2 engines in a similar arrangement to the S-IC, also using the outer engines for control. The S-II was 81.6 feet (24.87 m) tall with a diameter of 33 feet (10 m), identical to the S-IC, and thus was the largest cryogenic stage until the launch of the Space Shuttle in 1981. The S-II had a dry weight of about 80,000 pounds (36,000 kg); when fully fueled, it weighed 1,060,000 pounds (480,000 kg). The second stage accelerated the Saturn V through the upper atmosphere with 1,100,000 pounds-force (4,900 KN) of thrust in a vacuum.

When loaded, significantly more than 90 percent of the mass of the stage was propellant; however, the ultra-lightweight design had led to two failures in structural testing. Instead of having an intertank structure to separate the two fuel tanks as was done in the S-IC, the S-II used a common bulkhead that was constructed from both the top of the LOX tank and bottom of the LH2 tank. It consisted of two aluminum sheets separated by a honeycomb structure made of phenolic resin. This bulkhead had to insulate against the 126 °F (52 °C) temperature difference between the two tanks. The use of a common bulkhead saved 7,900 pounds (3.6 t) by both eliminating one bulkhead and reducing the stage's length. Like the S-IC, the S-II was transported from its manufacturing plant to the Cape by sea.

S-IVB third stage

The S-IVB was built by the Douglas Aircraft Company at Huntington Beach, California. It had one J-2 engine and used the same fuel as the S-II. The S-IVB used a common bulkhead to separate the two tanks. It was 58.6 feet (17.86 m) tall with a diameter of 21.7 feet (6.604 m) and was also designed with high mass efficiency, though not quite as aggressively as the S-II. The S-IVB had a dry weight of about 23,000 pounds (10,000 kg) and, fully fueled, weighed about 262,000 pounds (119,000 kg).

The S-IVB was the only rocket stage of the Saturn V small enough to be transported by Aero Spacelines Pregnant Guppy.

4.3 The Russian Soyuz Rockets

Soyuz (meaning "union", GRAU index 11A511) is a family of Soviet expendable launch systems developed by OKB-1 and manufactured by Progress Rocket Space Centre in Samara, Russia. With over 1700 flights since its debut in 1966, the Soyuz is the most frequently used launch vehicle in the world.

For nearly a decade between the final flight of the U.S. Space Shuttle program in 2011 and the 2020 first crewed mission on SpaceX's Falcon 9 rocket, Soyuz rockets were the only launch vehicles able and approved to transport astronauts to the International Space Station.

The Soyuz vehicles are used as the launcher for the crewed Soyuz spacecraft as part of the Soyuz program, as well as to launch uncrewed Progress supply spacecraft to the International Space Station and for commercial launches marketed and operated by Starsem and Arianespace. All Soyuz rockets use RP-1 and liquid oxygen (LOX) propellant, with the exception of the Soyuz-U2, which used Syntin, a variant of RP-1, with LOX. The Soyuz family is a subset of the R-7 family.

The Soyuz Spacecraft

Soyuz is a series of spacecraft designed for the Soviet space program by the Korolev Design Bureau (now RKK Energia) in the 1960s that remains in service today, having made more than 140 flights. The Soyuz succeeded the Voskhod spacecraft and was originally built as part of the Soviet crewed lunar programs. The Soyuz spacecraft is launched on a Soyuz rocket, the most reliable launch vehicle in the world to date. The Soyuz rocket design is

Types of Rocket Propulsion and Potential Space Drives

based on the Vostok launcher, which in turn was based on the 8K74 or R-7A Semyorka, a Soviet intercontinental ballistic missile. All Soyuz spacecraft are launched from the Baikonur Cosmodrome in Kazakhstan. After the retirement of the Space Shuttle in 2011, the Soyuz served as the only means for Americans to make crewed space flights until the first flight of VSS Unity in 2018, and the only means for Americans to reach the International Space Station until the first flight of Dragon 2 Crew variant on May 30, 2020. The Soyuz is heavily used in the ISS program.

Types of Rocket Propulsion and Potential Space Drives

4.4 The Space Shuttle

The Space Shuttle was a partially reusable low Earth orbital spacecraft system that was operated from 1981 to 2011 by the National Aeronautics and Space Administration (NASA) as part of the Space Shuttle program. Its official program name was Space Transportation System (STS), taken from a 1969 plan for a system of reusable spacecraft of which it was the only item funded for development. The first of four orbital test flights occurred in 1981, leading to operational flights beginning in 1982. Five complete Space Shuttle orbiter vehicles were built and flown on a total of 135 missions from 1981 to 2011, launched from the Kennedy Space Center (KSC) in Florida.

Operational missions launched numerous satellites, Interplanetary probes, and the Hubble Space Telescope (HST); conducted science experiments in orbit; and participated in construction and servicing of the International Space Station. The Space Shuttle fleet's total mission time was 1322 days, 19 hours, 21 minutes and 23 seconds.

Space Shuttle components include the Orbiter Vehicle (OV) with three clustered Rocketdyne RS-25 main engines, a pair of recoverable solid rocket boosters (SRBs), and the expendable external tank (ET) containing liquid hydrogen and liquid oxygen. The Space Shuttle was launched vertically, like a conventional rocket, with the two SRBs operating in parallel with the orbiter's three main engines, which were fueled from the ET. The SRBs were jettisoned before the vehicle reached orbit, and the ET was jettisoned just before orbit insertion, which used the orbiter's two Orbital Maneuvering System (OMS) engines.

At the conclusion of the mission, the orbiter fired its OMS to deorbit and reenter the atmosphere. The orbiter was protected during reentry by its thermal protection system tiles, and it glided as a spaceplane to a runway landing, usually to the Shuttle Landing Facility at KSC, Florida, or to Rogers Dry Lake in Edwards Air Force Base, California. If the landing occurred at Edwards, the orbiter was flown back to the KSC on the Shuttle Carrier Aircraft, a specially modified Boeing 747.

The Space Shuttle used a combination of engines from liquid fueled in the launch stages and on the orbiter, to the two large solid rocket boosters.

Types of Rocket Propulsion and Potential Space Drives

Solid Rocket Boosters

The Space Shuttle Solid Rocket Booster (Space Shuttle SRB) was the first solid-propellant rocket to be used for primary propulsion on a vehicle used for human spaceflight and provided the majority of the Space Shuttle's thrust during the first two minutes of flight. After burnout, they were jettisoned and parachuted into the Atlantic Ocean where they were recovered, examined, refurbished, and reused.

The Space Shuttle SRB was the most powerful solid rocket motor ever flown. Each provided a maximum 14.7 MN (3,300,000 lbf) thrust, roughly double the most powerful single-combustion chamber liquid-propellant rocket engine ever flown, the Rocketdyne F-1. With a combined mass of about 1,180 t (1,160 long tons; 1,300 short tons), they comprised over half the mass of the Shuttle stack at liftoff. The motor segments of the SRBs were manufactured by Thiokol of Brigham City, Utah, which was later purchased by ATK. The prime contractor for most other components of the SRBs, as well as for the integration of all the components and retrieval of the spent SRBs, was USBI, a subsidiary of Pratt and Whitney. This contract was subsequently transitioned to United Space Alliance, a limited liability company joint venture of Boeing and Lockheed Martin.

Out of 270 SRBs launched over the Shuttle program, all but four were recovered – those from STS-4 (due to a parachute malfunction) and STS-51-L (Challenger disaster). Over 5,000 parts were refurbished for reuse after each flight. The final set of SRBs that launched STS-135 included parts that flew on 59 previous missions, including STS-1. Recovery also allowed post-flight examination of the boosters, identification of anomalies, and incremental design improvements.

4.5 The Russian RD-180 Engine

The RD-180 is a good example of Soviet Engine technology which outperforms anything developed by the West in similar timeframes. This is why it was purchased to use in American rockets for years.

The roots of the RD-180 rocket engine extend back into the Soviet Energia launch vehicle project. The RD-170, a four-chamber engine, was developed for use on the strap-on boosters for this vehicle, which ultimately was used to lift the Buran orbiter. This engine was scaled down to a two-chamber version by combining the RD-170's combustion devices with half-size turbomachinery. After successful performances in engine tests on a test stand and high-level agreements between the US government and the Russian government, the engines were imported to the US for use on the Lockheed Martin Atlas III, with first flight in 2000. The engine is also used on the United Launch Alliance Atlas V, the successor to the Atlas III.

The engine has similar design features to the NK-33, which was developed by a different bureau (Kuznetzov) nearly a decade earlier.

Design and specifications

The combustion chambers of the RD-180 share a single turbopump unit, much like in its predecessor, the four-chambered RD-170. The RD-180 is fueled by an RP-1/LOX mixture and uses an extremely efficient, high-pressure staged combustion cycle. The engine runs with an oxidizer-to-fuel ratio of 2.72 and employs an oxygen-rich preburner, unlike typical fuel-rich US designs. The thermodynamics of the cycle allow an oxygen-rich preburner to give a greater power-to-weight ratio, but with the drawback that high-pressure, high-temperature gaseous oxygen must be transported throughout the engine. The movements of the engine nozzles are controlled by four hydraulic actuators. The engine can be throttled from 47% to 100% of nominal thrust.

4.6   SpaceX Reusable Rockets

SpaceX has developed a revolutionary reusable rocket system. Here is an article describing this accomplishment from 2018:

> A few small glitches marred an otherwise stellar year as SpaceX honed its reusable rocket technology. AT THE BEGINNING of 2018, Elon Musk predicted that SpaceX would pull off 30 launches. The goal seemed far-fetched; among other reasons, some of those flights were planned for the Falcon Heavy, which at the time had yet to fly. Indeed, the company didn't hit that figure. But the 21 launches it did pull off in 2018 still amount to a staggering achievement for the 16-year-old company.
>
> Building off its earlier momentum—eight launches in 2016 and 18 in 2017—SpaceX's reusable rocket technology moved out of the proof-of-concept stage to become the backbone of a growing fleet of flight-proven rockets. Although the company landed its first rocket in 2015, it took until 2017 for SpaceX to reuse its

first booster. This year landings became almost routine, and engineers bid farewell to the moderately reusable Falcons of yesterday, ushering in an era of more capable Falcons, dubbed the Block 5.

The souped-up version of the company's flagship rocket has performed beautifully since lifting its first payload, a communications satellite for the country of Bangladesh. According to SpaceX, the modifications (which include improved engines, a more durable interstage, titanium grid fins, and a new thermal protection system) will enable each Block 5 to fly 10 times or more before needing light refurbishments, and up to 100 times before retirement. This year all but two of the 14 rockets that SpaceX attempted to land stuck their landings.

After that first flight in May, Musk announced that sometime next year, SpaceX would launch and land the same Block 5 booster twice within a 24-hour period. To that end, the company has shown that it was able reuse the same booster three times; it also opened a new landing site, which should help reduce post-launch processing times.

Another long-standing goal was to debut its heavy-lift rocket, the Falcon Heavy. First estimated to fly back in 2013, the Falcon Heavy took its inaugural flight on Feb. 6th. After sending a cherry red Tesla roadster (complete with Starman pilot) on a journey past Mars, the three cores of the Falcon Heavy—each essentially its own Falcon 9 rocket—returned to Earth. Two of the boosters touched down in perfect synchronization on LZ-1 and LZ-2, SpaceX's designated landing zones at Cape Canaveral. The Heavy's center booster, however, missed its targeted landing spot on a waiting drone ship, and plopped into the Atlantic Ocean.

That was it for the Falcon Heavy in 2018; its next two flights were delayed to sometime next year. Still, that one test flight paid off for SpaceX. In a surprising move, the Air Force not only certified the Falcon Heavy for military payloads but also awarded the heavy-lifter its first major contract: the AFSPC-52 mission. Valued at $130 million, that contract will see SpaceX deposit an Air Force satellite in space sometime around 2020.

Not every negotiation went its way. SpaceX missed out on a slice of a government contract worth $2 billion to build hardware that could eventually launch national security payloads. The money ended up going to three of its competitors, ULA, Blue Origin, and Northrop Grumman.

The company also pulled back from some of its targets. During a press briefing for the Falcon Heavy flight, Elon Musk revealed that the company's plans to send passengers around the moon on the Falcon Heavy would be scrapped. So what happened to the people who booked tickets? That answer wouldn't be revealed until months later, when Musk announced that Yusaku Maezawa was the mysterious billionaire who had reserved both seats on the Falcon Heavy in the first place. Now Maezawa was slated to be the first paying customer on SpaceX's next-generation rocket—the BFR (or Big Falcon Rocket). Billed from the start as an interplanetary transport vehicle, BFR (which is composed of two parts, a rocket and a spaceship capable of carrying hundreds of people into space) is perhaps more suited to this mission than the Falcon

Types of Rocket Propulsion and Potential Space Drives

Heavy, allowing the heavy-lifter to rake in more sweet government contracts.

Types of Rocket Propulsion and Potential Space Drives

5.0 Existing and Future Engine Designs

One of the objectives of space travel in our Solar System is to design new rockets with a higher specific impulse and which can run continuously for weeks or months.

A nuclear rocket which could do this could reduce travel times to Mars from months to weeks or days.

5.1 Nerva

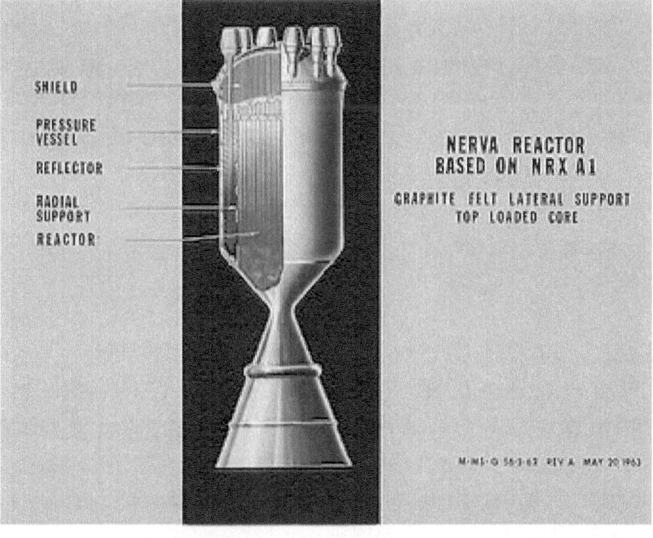

SHIELD

PRESSURE
VESSEL

REFLECTOR

RADIAL
SUPPORT

REACTOR

NERVA REACTOR
BASED ON NRX A1

GRAPHITE FELT LATERAL SUPPORT
TOP LOADED CORE

M-MS-G 58-3-63 REV A MAY 20 1963

The Nuclear Engine for Rocket Vehicle Application (NERVA) was a nuclear thermal rocket engine development program that ran for roughly two decades. Its principal objective was to "establish a technology base for nuclear rocket engine systems to be utilized in the design and development of propulsion systems for space mission application" NERVA was a joint effort of the Atomic Energy Commission (AEC) and the National Aeronautics and

Space Administration (NASA), and was managed by the Space Nuclear Propulsion Office (SNPO) until the program ended in January 1973. SNPO was led by NASA's Harold Finger and AEC's Milton Klein.

NERVA had its origins in Project Rover, an AEC research project at the Los Alamos Scientific Laboratory (LASL) with the initial aim of providing a nuclear-powered upper stage for the United States Air Force intercontinental ballistic missiles, which are more powerful than chemical engines. After the formation of NASA in 1958, Project Rover was continued as a civilian project and was reoriented to producing a nuclear powered upper stage for NASA's Saturn V Moon rocket. Reactors were tested at very low power before being shipped to Jackass Flats in the Nevada Test Site. While LASL concentrated on reactor development. NASA built and tested complete rocket engines.

The AEC, SNPO, and NASA considered NERVA to be a highly successful program in that it met or exceeded its program goals. NERVA demonstrated that nuclear thermal rocket engines were a feasible and reliable tool for space exploration, and at the end of 1968 SNPO certified that the latest NERVA engine, the XE, met the requirements for a human mission to Mars. It had strong political support from Senators Clinton P. Anderson and Margaret Chase Smith but was cancelled by President Richard Nixon in 1973. Although NERVA engines were built and tested as much as possible with flight-certified components and the engine was deemed ready for integration into a spacecraft, they never flew in space. Plans for deep space exploration generally require the power of nuclear rocket engines, and all spacecraft concepts featuring them use derivative designs from the NERVA.

Types of Rocket Propulsion and Potential Space Drives

5.2 SpaceX Engine Designs

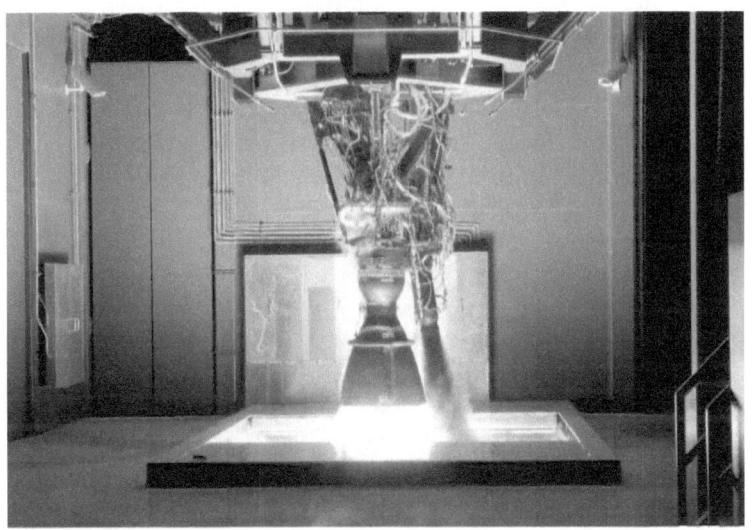

Kerosene-based engines

SpaceX has developed two kerosene-based engines through 2013, the Merlin 1 and Kestrel, and has publicly discussed a much larger concept engine high-level design named Merlin 2. Merlin 1 powered the first stage of the Falcon 1 launch vehicle and is used both on the first and second stages of the Falcon 9 and Falcon Heavy launch vehicles. The Falcon 1 second stage was powered by a Kestrel engine.

Merlin 1

Merlin 1 is a family of LOX/RP-1 rocket engines developed 2003–2012. Merlin 1A and Merlin 1B utilized an ablatively cooled carbon fiber composite nozzle. Merlin 1A produced 340 kilonewtons (76,000 lbf) of thrust and was used to power the first stage of the first two Falcon 1 flights in 2006 and 2007. Merlin 1B had a somewhat more powerful turbo-pump, and generated more thrust, but was never flown on a flight vehicle before SpaceX's move to the Merlin 1C.

The Merlin 1C was the first in the family to use a regeneratively cooled nozzle and combustion chamber. It was first fired with a full mission duty firing in 2007, first flew on the third Falcon 1 mission in August 2008, powered the "first privately-developed liquid-fueled rocket to successfully reach orbit" (Falcon 1 Flight 4) in September 2008, and subsequently powered the first five Falcon 9 flights — each flown with a version 1.0 Falcon 9 launch vehicle — from 2010 through 2013.

The Merlin 1D, developed in 2011–2012 also has a regeneratively cooled nozzle and combustion chamber. It has a vacuum thrust of 690 kN (155,000 lbf), a vacuum specific impulse (Isp) of 310 s, an increased expansion ratio of 16 (as opposed to the previous 14.5 of the Merlin 1C) and chamber pressure of 9.7 MPa (1,410 psi). A new feature for the engine is the ability to throttle from 100% to 70%. The engine's 150:1 thrust-to-weight ratio is the highest ever achieved for a rocket engine. The first flight of the Merlin 1D engine was also the maiden Falcon 9 v1.1 flight. On 29 September 2013, the Falcon 9 Flight 6 mission successfully launched the Canadian Space Agency's CASSIOPE satellite into polar orbit, and proved that the Merlin 1D could be restarted to control the first stage's re-entry back into the atmosphere—part of the SpaceX reusable launch system flight test program—a necessary step in making the rocket reusable.

Kestrel

Kestrel was a LOX/RP-1 pressure-fed rocket engine, and was developed by SpaceX as the Falcon 1 rocket's second stage main engine. It was built around the same pintle architecture as SpaceX's Merlin engine but does not have a turbo-pump, and is fed only by tank pressure. Its nozzle was ablatively cooled in the chamber and radiatively cooled in the throat, and is fabricated from a high strength niobium alloy. Thrust vector control is provided by electro-mechanical actuators on the engine dome for pitch and yaw. Roll control – and attitude control during the coast phase – is provided by helium cold gas thrusters.

Methane-based engines

In November 2012, methalox engines came on the scene when SpaceX CEO Elon Musk announced a new direction for propulsion side of the company: developing methane/LOX rocket engines. SpaceX work on methane/LOX (methalox) engines is strictly to support the company's Mars technology development program. They had no plans to build an upper stage engine for the Falcon 9 or Falcon Heavy using methalox propellant. However, on November 7, 2018 Elon Musk tweeted, "Falcon 9 second stage will be upgraded to be like a mini-BFR Ship," which may imply the use of a Raptor engine on this new second stage. The focus of the new engine development program is exclusively on the full-size Raptor engine for the Mars-focused mission.

Raptor

Raptor is a family of methane/liquid oxygen rocket engines under development by SpaceX since the late 2000s, although LH2/LOX propellant mix was originally under study when the Raptor concept development work began in 2009. When first mentioned by SpaceX in 2009, the term "Raptor" was applied exclusively to an upper stage engine concept. SpaceX discussed in October 2013 that they intended to build a family of methane-based Raptor rocket engines, initially announcing that the engine would achieve 2.94 meganewtons (661,000 lbf) vacuum thrust. In February 2014, they announced that the Raptor engine would be used on the Mars Colonial Transporter. The booster would utilize multiple Raptor engines, similar to the use of nine Merlin 1s on each Falcon 9 booster core. The following month, SpaceX confirmed that as of March 2014, all Raptor development work is exclusively on this single very large rocket engine, and that no smaller Raptor engines were in the current development mix.

The Raptor methane/LOX engine uses a highly efficient and theoretically more reliable full-flow staged combustion cycle, a departure from the open gas generator cycle system and LOX/kerosene propellants used on the current Merlin 1 engine series. As of February 2014, preliminary designs of Raptor were looking at producing 4.4 meganewtons (1,000,000 lbf) of thrust with a vacuum specific impulse (Isp) of 363 seconds (3.56 km/s) and a sea-level Isp of 321 seconds (3.15 km/s), although later concept sizes being looked at were closer to 2.2 MN (500,000 lbf).

Initial component-level testing of Raptor technology began in May 2014, with an injector element test. The first complete Raptor development engine, approximately one-third the size of the full-scale engines planned for the use on various parts of the Starship, with approximately 1,000 kN (220,000 lbf) thrust, began testing on a ground test stand in September 2016. The test nozzle has an expansion ratio of only 150, in order to eliminate flow separation problems while tested in Earth's atmosphere.

Raptor's full-flow staged combustion cycle will pass 100 percent of the oxidizer (with a low-fuel ratio) to power the oxygen turbine pump, and 100 percent of the fuel (with a low-oxygen ratio) to power the methane turbine pump. Both streams—oxidizer and fuel—will be completely in the gas phase before they enter the combustion chamber.

Prior to 2016, only two full-flow staged combustion rocket engines had ever progressed sufficiently to be tested on test stands: the Soviet RD-270 project in the 1960s and the Aerojet Rocketdyne Integrated powerhead demonstration project in the mid-2000s, which did not test a complete engine but rather only the powerhead.

Other characteristics of the full-flow design are projected to further increase performance or reliability, with the possibility to do design trade-offs of one against the other.

Types of Rocket Propulsion and Potential Space Drives

Types of Rocket Propulsion and Potential Space Drives

5.3 Ion Drives

An ion thruster or ion drive is a form of electric propulsion used for spacecraft propulsion. It creates thrust by accelerating ions using electricity.

An ion thruster ionizes a neutral gas by extracting some electrons out of atoms, creating a cloud of positive ions. These ion thrusters rely mainly on electrostatics as ions are accelerated by the Coulomb force along an electric field. Temporarily stored electrons are finally reinjected by a neutralizer in the cloud of ions after it has passed through the electrostatic grid, so the gas becomes neutral again and can freely disperse in space without any further electrical interaction with the thruster. Electromagnetic thrusters on the contrary use the Lorentz force to accelerate all species (free electrons as well as positive and negative ions) in the same direction whatever their electric charge, and are specifically referred to as plasma propulsion engines, where the electric field is not in the direction of the acceleration.

Ion thrusters in operational use have an input power need of 1–7 kW (1.3–9.4 hp), exhaust velocity 20–50 km/s

Types of Rocket Propulsion and Potential Space Drives

(45,000–112,000 mph), thrust 25–250 millinewtons (0.090–0.899 ozf) and efficiency 65–80% though experimental versions have achieved 100 kilowatts (130 hp), 5 newtons (1.1 lbf).

The Deep Space 1 spacecraft, powered by an ion thruster, changed velocity by 4.3 km/s (9,600 mph) while consuming less than 74 kg (163 lb) of xenon. The Dawn spacecraft broke the record, with a velocity change of 11.5 km/s (26,000 mph).

Applications include control of the orientation and position of orbiting satellites (some satellites have dozens of low-power ion thrusters) and use as a main propulsion engine for low-mass robotic space vehicles (such as Deep Space 1 and Dawn).

Ion thrust engines are practical only in the vacuum of space and cannot take vehicles through the atmosphere because ion engines do not work in the presence of ions outside the engine. Additionally, the engine's minuscule thrust cannot overcome any significant air resistance. Spacecraft rely on conventional chemical rockets to reach their initial orbit.

5.4 Project Orion

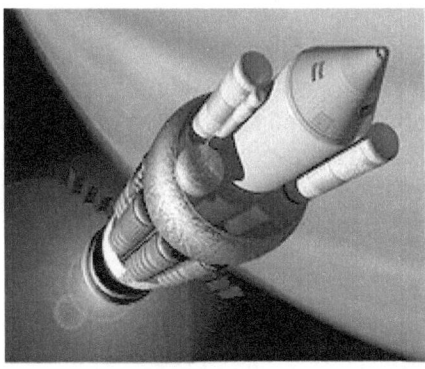

Project Orion was a study of a spacecraft intended to be directly propelled by a series of explosions of atomic bombs behind the craft (nuclear pulse propulsion). Early versions of this vehicle were proposed to take off from the ground (with significant associated nuclear fallout); later versions were presented for use only in space. Six non-nuclear tests were conducted using models. The project was eventually abandoned for multiple reasons such as the Partial Test Ban Treaty which banned nuclear explosions in space as well as concerns over nuclear fallout.

The idea of rocket propulsion by combustion of explosive substance was first proposed by Russian explosives expert Nikolai Kibalchich in 1881, and in 1891 similar ideas were developed independently by German engineer Hermann Ganswindt. Robert A. Heinlein mentions powering spaceships with nuclear bombs in his 1940 short story "Blowups Happen." Real life proposals of nuclear propulsion were first made by Stanislaw Ulam in 1946, and preliminary calculations were made by F. Reines and Ulam in a Los Alamos memorandum dated 1947. The actual project, initiated in 1958, was led by Ted Taylor at General

Types of Rocket Propulsion and Potential Space Drives

Atomics and physicist Freeman Dyson, who at Taylor's request took a year away from the Institute for Advanced Study in Princeton to work on the project.

The Orion concept offered high thrust and high specific impulse, or propellant efficiency, at the same time. The unprecedented extreme power requirements for doing so would be met by nuclear explosions, of such power relative to the vehicle's mass as to be survived only by using external detonations without attempting to contain them in internal structures. As a qualitative comparison, traditional chemical rockets—such as the Saturn V that took the Apollo program to the Moon—produce high thrust with low specific impulse, whereas electric ion engines produce a small amount of thrust very efficiently. Orion would have offered performance greater than the most advanced conventional or nuclear rocket engines then under consideration. Supporters of Project Orion felt that it had potential for cheap interplanetary travel, but it lost political approval over concerns with fallout from its propulsion

The Partial Test Ban Treaty of 1963 is generally acknowledged to have ended the project. However, from Project Longshot to Project Daedalus, Mini-Mag Orion, and other proposals which reach engineering analysis at the level of considering thermal power dissipation, the principle of external nuclear pulse propulsion to maximize survivable power has remained common among serious concepts for interstellar flight without external power beaming and for very high-performance interplanetary flight. Such later proposals have tended to modify the basic principle by envisioning equipment driving detonation of much smaller fission or fusion pellets, in contrast to Project Orion's larger nuclear pulse units (full nuclear bombs) based on less speculative technology

Types of Rocket Propulsion and Potential Space Drives

6.0 Laser Spacecraft Propulsion

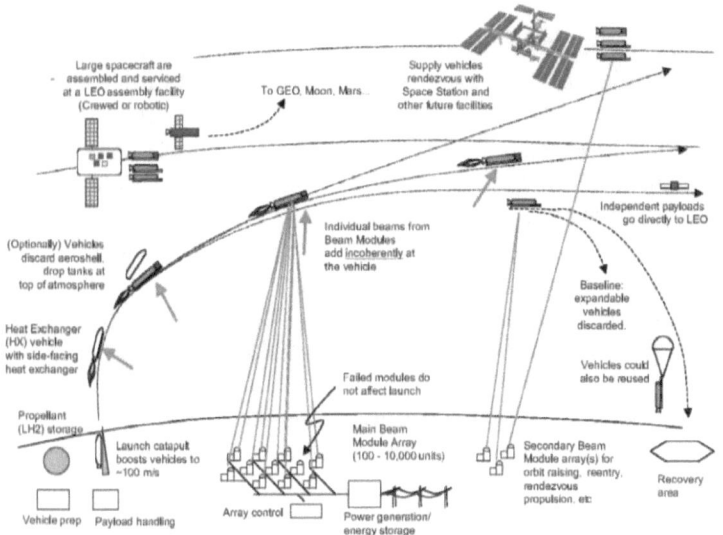

Figure 1: Laser Launch Architecture With Modular Ground-Based Laser Array

Laser propulsion is a totally different approach than rocket engines. Instead of an engine which pushes the rocket from its own force, a laser LightSail would be pushed by an array of lasers on or near the Earth.

A laser-pushed LightSail is a thin reflective sail similar to a solar sail, in which the sail is being pushed by a laser, rather than the sun. The advantage of LightSail propulsion is that the vehicle does not carry either the energy source or the reaction mass for propulsion, and hence the limitations of the Tsiolkovsky rocket equation to achieving high velocities are avoided. Use of a laser-pushed LightSail was proposed initially by Marx in 1966, as a method of Interstellar travel that would avoid extremely high mass ratios by not carrying fuel, and analyzed in detail by physicist Robert L. Forward in 1989. Further analysis of the concept was done by Landis, Mallove and Matloff, Andrews and others.

The beam has to have a large diameter so that only a small portion of the beam misses the sail due to diffraction and the laser or microwave antenna has to have a good pointing stability so that the craft can tilt its sails fast enough to follow the center of the beam. This gets more important when going from interplanetary travel to interstellar travel, and when going from a fly-by mission, to a landing mission, to a return mission. The laser may alternatively be a large phased array of small devices, which get their energy directly from solar radiation.

The laser-pushed sail is proposed as a method of propelling a small interstellar probe by the Breakthrough Starshot project.

Another method of moving a much larger spacecraft to high velocities is by using a laser system to propel a stream of much smaller sails. Each alternative mini sail is slowed down by a laser from the home system so that they collide at ionizing velocities. The ionizing collisions could then be used to interact with a powerful magnetic field on the spacecraft to provide a force to power and move it. An extension of the idea is to have nuclear materials on the mini sails that undergo fission or fusion to provide a much more powerful force but the collision velocities would have to be much higher.

Types of Rocket Propulsion and Potential Space Drives

7.0 New Physics & Faster than Light Drives

The question as to whether Aliens and UFOs have visited the Earth is a major and exciting source of speculation. I've even written a couple of books on the subject and my research and other reading convinced me that these other races of beings have visited Earth in the past.

If this is true, then they must have also had some type of much faster spacecraft including space drives which can exceed the speed of light.

Here is some of the current thinking about novel drive technologies which humans might use to reach the stars in our own lifetimes.

7.1 EmDrive

Rocket engines operate by expelling propellant, which acts as a reaction mass and which produces thrust per Newton's third law of motion. In the 1960s, extensive research was conducted on two designs which emit high-velocity ionized gases in similar ways: ion thrusters that convert propellant to ions and accelerate and eject them via electric potentials, and plasma thrusters that convert propellant to plasma ions and accelerate and eject them via plasma currents. All designs for electromagnetic propulsion operate on the principle of reaction mass.

A drive which does not expel propellant in order to produce a reaction force, providing thrust while being a closed system with no external interaction, would be a reactionless drive. Such a drive would violate the conservation of momentum and Newton's third law, leading many physicists to consider the idea pseudoscience. Such drives are a popular concept in science fiction, and their implausibility contributes to enthusiasm for exploring such designs.

The original proposal for an RF resonant cavity thruster came from Roger Shawyer in 2001. He proposed a design with a conical cavity, which he called "EmDrive". He claimed that it produced thrust in the direction of the base of the cavity. Guido Fetta later built the Cannae Drive based on Shawyer's concept as a resonant thruster with a pillbox-shaped cavity. Since 2008, a few physicists have tested their own models, trying to reproduce the results claimed by Shawyer and Fetta. Juan Yang at Xi'an's Northwestern Polytechnical University (NWPU) was unable to reproducibly measure thrust from their models, over the course of 4 years.

In 2016, Harold White's group at NASA's Advanced Propulsion Physics Laboratory reported in the Journal of Propulsion and Power that a test of their own model had observed a small thrust. In December 2016, Yue Chen of the communication satellite division of the China Academy of Space Technology (CAST), said his team had tested several prototypes, observed thrust, and was carrying out in-orbit verification. In September 2017, Chen talked about this CAST project again in an interview on CCTV.

Media coverage of experiments using these designs has been controversial and polarized. The EmDrive first drew attention, both credulous and dismissive, when New Scientist wrote about it as an "impossible" drive in 2006. Media outlets were later criticized for misleading claims that a resonant cavity thruster had been "validated by NASA" following White's first tentative test reports in 2014. Scientists have continued to note the lack of unbiased coverage, from both polarized sides.

In 2006, responding to the New Scientist piece, mathematical physicist John C. Baez at the University of California, Riverside, and Australian science-fiction writer Greg Egan, said the positive results reported by Shawyer were likely misinterpretations of experimental errors.

In 2014, White's conference paper suggested that resonant cavity thrusters could work by transferring momentum to the "quantum vacuum virtual plasma. "Baez and Carroll criticized this explanation, because in the standard description of vacuum fluctuations, virtual particles do not behave as a plasma; Carroll also noted that the quantum vacuum has no "rest frame", providing nothing to push against, so it can't be used for propulsion. In the same way, physicists James F. Woodward and Heidi Fearn published two papers showing that electron-positron virtual pairs of the quantum vacuum, discussed by White as a potential virtual plasma propellant, could not account for thrust in any isolated, closed electromagnetic system such as a quantum vacuum thruster.

Types of Rocket Propulsion and Potential Space Drives

7.2 Alcubierrre Drive

The Alcubierre drive, Alcubierre warp drive, or Alcubierre metric (referring to metric tensor) is a speculative idea based on a solution of Einstein's field equations in general relativity as proposed by theoretical physicist Miguel Alcubierre, by which a spacecraft could achieve apparent faster-than-light travel if a configurable energy-density field lower than that of vacuum (that is, negative mass) could be created.

Rather than exceeding the speed of light within a local reference frame, a spacecraft would traverse distances by contracting space in front of it and expanding space behind it, resulting in effective faster-than-light travel. Objects cannot accelerate to the speed of light within normal space-time; instead, the Alcubierre drive shifts space around an object so that the object would arrive at its destination faster than light would in normal space without breaking any physical laws.

Although the metric proposed by Alcubierre is consistent with the Einstein field equations, construction of such a drive is not necessarily possible. The proposed mechanism of the Alcubierre drive implies a negative energy density and therefore requires exotic matter. So if exotic matter with the correct properties cannot exist, then the drive could not be constructed. At the close of his original article, however, Alcubierre argued (following an argument developed by physicists analyzing traversable wormholes) that the Casimir vacuum between parallel plates could fulfill the negative-energy requirement for the Alcubierre drive.

Another possible issue is that, although the Alcubierre metric is consistent with Einstein's equations, general relativity does not incorporate quantum mechanics. Some physicists have presented arguments to suggest that a theory of quantum gravity (which would incorporate both theories) would eliminate those solutions in general relativity that allow for backwards time travel (see the chronology protection conjecture) and thus make the Alcubierre drive invalid.

7.3 Worm Hole Space Drive

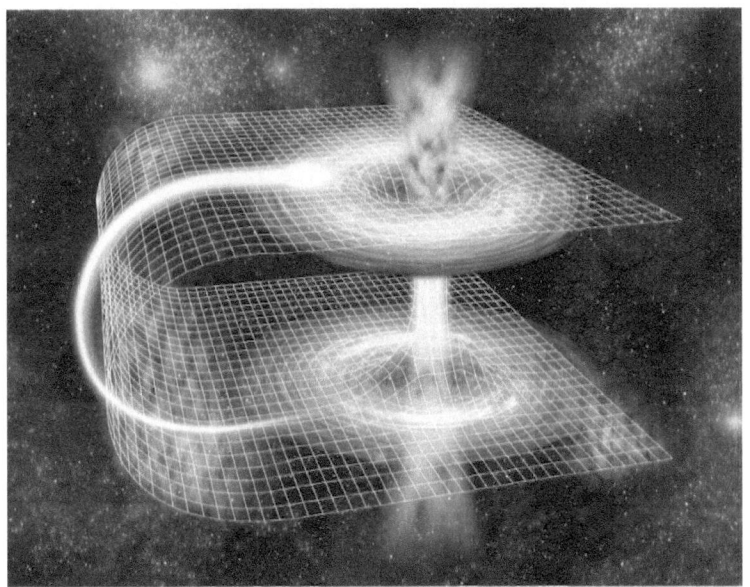

Also known as wormholes, Einstein-Rosen bridges are perhaps the most commonly known means of interstellar travel — and the most likely to actually exist. Albert Einstein's general theory of relativity predicted wormholes, although we haven't seen one yet.

An Einstein-Rosen bridge, to put it as simply as possible, is a shortcut through space caused by the warping of space-time. Massive objects like stars or black holes bend time and space like a bowling ball on a trampoline. A massive enough object could bend space-time to create a connection between two otherwise distinct points.

Or, as one character says to another in Interstellar, picture a piece of paper folded on itself, then pierced with a pencil. If you want to travel across the surface of the paper to get point to point, it'll take a while. But if you can use the pencil as a bridge, it's a significantly shorter trip.

The entrance to a wormhole has often been represented as a tube, which makes sense given the name. But it's also inaccurate. Interstellar, in one of its most intense scenes, got it right. From our perspective in 3-D space, a wormhole should look like a sphere.

Wormholes are an attractive approach to FTL technology because they don't require you to break the speed of light. Physics tells us nothing can go faster than light. But with wormholes — shortcuts, basically — spacecraft could enter and exit at sub-light speeds.

That makes them a perfect fit for any fictional plot, allowing a quick way to get from Point A to Point B. In Douglas Adams' satirical Hitchhiker's Guide, the Earth was demolished by the Vogons to make way for a hyperspace bypass — a wormhole by any other name — which would supposedly speed up galactic travel.

Wormhole theory also allows for the possibility of traveling between universes. Einstein-Rosen bridges are key to Rick and Morty's various misadventures; the mad scientist's portal gun can open doorways to other galaxies and parallel realities.

Some spoilsport scientists claim a wormhole caused by some sort of supermassive black hole would likely be too unstable to go through. A 1962 paper by John Archibald Wheeler and Robert W. Fuller argued such a bridge would collapse too quickly. On the other hand, physicists like Stephen Hawking and Kip Thorne have theorized wormholes could be theoretically stabilized with the right amount of energy. But that's a big "if."

8.0 Summary

Space rocket and spaceship activities are at a high water mark and the greatest level of activity in the last fifty years.

The good news is that there is a lot of development going on for new rockets, engines, and methods of propulsion.

Engineers and Scientists recognize that we will need to develop higher thrust engines and one with a larger specific impulse to address many of the desired trips to Mars and other planets. These better rockets would also be useful to have for our shorter trips to the Moon and back.

Martin K. Ettington

August 2020

Types of Rocket Propulsion and Potential Space Drives

Types of Rocket Propulsion and Potential Space Drives

9.0 Bibliography

1. Rocket Englines.
https://en.wikipedia.org/wiki/Rocket_engine. [Online]

2. Alcubierre Drive.
https://en.wikipedia.org/wiki/Alcubierre_drive. [Online]

3. RD-180. https://en.wikipedia.org/wiki/RD-180. [Online]

4. The Soyuz Rocket Family.
https://en.wikipedia.org/wiki/Soyuz_(rocket_family).
[Online]

5. Specific Impulse.
https://en.wikipedia.org/wiki/Specific_impulse. [Online]

6. Nuclear Thermal Rocket.
https://en.wikipedia.org/wiki/Nuclear_thermal_rocket.
[Online]

7. NERVA. https://en.wikipedia.org/wiki/NERVA. [Online]

8. Project Orion.
https://en.wikipedia.org/wiki/Project_Orion_(nuclear_propul
sion). [Online]

9. Project Orion.
https://en.wikipedia.org/wiki/Project_Orion_(nuclear_propul
sion). [Online]

10. Ion Thruster. https://en.wikipedia.org/wiki/Ion_thruster.
[Online]

11. SAturn V. https://en.wikipedia.org/wiki/Saturn_V.
[Online]

12. Soyuz Spacecraft.
https://en.wikipedia.org/wiki/Soyuz_(spacecraft). [Online]

13. Space Shuttle.
https://en.wikipedia.org/wiki/Space_Shuttle. [Online]

14. V2 Rocket. https://en.wikipedia.org/wiki/V-2_rocket.
[Online]

15. Laser Propulsion.
https://en.wikipedia.org/wiki/Laser_propulsion. [Online]

16. Resonant Cavity Thruster.
https://en.wikipedia.org/wiki/RF_resonant_cavity_thruster.
[Online]

17. Faster than light space interstellar travel.
https://mashable.com/feature/faster-than-light-space-
interstellar-travel/. [Online]

18. Spacex rocket engines.
https://en.wikipedia.org/wiki/SpaceX_rocket_engines.
[Online]

19. the year spacex made us all believe in reusable
rockets. https://www.wired.com/story/this-year-spacex-
made-us-all-believe-in-reusable-rockets/. [Online]

20. Space Shuttle Solid Rocket Booster.
https://en.wikipedia.org/wiki/Space_Shuttle_Solid_Rocket_
Booster. [Online]

21. Soyuz Spacecraft.
https://en.wikipedia.org/wiki/Soyuz_(spacecraft). [Online]

10.0   Index